2

ISAAC ASIMOV'S
Library of the Universe

The Earth's Moon

by Isaac Asimov

Gareth Stevens Publishing
Milwaukee

Library of Congress Cataloging-in-Publication Data

Asimov, Isaac, 1920-
   The Earth's Moon.

   (Isaac Asimov's library of the universe)
   Bibliography: p.
   Includes index.
   Summary: Examines the many facets and puzzles of our Moon, including its phases and eclipses, its early discoveries and modern exploration, and its possible origins and future prospects.
   1. Moon—Juvenile literature. [1. Moon] I. Title. II. Asimov, Isaac, 1920- Library of the Universe.
QB582.A85 1988   523.3   87-42601
ISBN 1-55532-382-0
ISBN 1-55532-357-X (lib. bdg.)

A Gareth Stevens Children's Books edition. Edited, designed, and produced by

Gareth Stevens, Inc.
7317 West Green Tree Road   Milwaukee, Wisconsin 53223, USA

Cover photograph © Frank Zullo 1985

Designer: Laurie Shock.
Picture research: Kathy Keller.
Artwork commissioning: Kathy Keller and Laurie Shock.
Project editor: Mark Sachner.

Technical adviser and consulting editor: Greg Walz-Chojnacki.

1 2 3 4 5 6 7 8 9 93 92 91 90 89 88

CONTENTS

Introduction

The Universe we live in is an enormously large place. Only in the last 50 years or so have we learned how large it really is.

It's only natural that we would want to understand the place we live in, so in the last 50 years we have developed new instruments to help us learn about it. We have probes, satellites, radio telescopes, and many other things that tell us far more about the Universe than could possibly be imagined when I was young.

Nowadays, we have seen planets close up, all the way to distant Uranus. We have mapped Venus through its clouds. We have seen dead volcanoes on Mars and live ones on Io, one of Jupiter's satellites. We have detected strange objects no one knew anything about till recently: quasars, pulsars, black holes. We have learned amazing facts about how the Universe was born, and we have some ideas about how it may die. Nothing can be more astonishing and more interesting.

But the nearest object of all is our Moon. It is only a quarter of a million miles (400,000 km) away. The next nearest object, which is the planet Venus, is 100 times as far away. Mars is 150 times as far away, and everything else is much, much farther. In fact, the Moon is only three days away by rocket ship, and it is the only world other than Earth that human beings have stood on.

Let's learn a little about the Moon.

A daguerreotype (an early form of photograph) of the Moon made on February 26, 1852. This is one of the first pictures taken of the Moon.

Earth's Neighbor

There is no doubt about it: The Moon is the ruler of our night sky. Everything else in the night sky is just a point of light. But the Moon is large enough and close enough to give us light at night. It is close enough for its gravitational pull to drag the sea upward and cause the tides. We can see both shadows and bright spots on the Moon's surface. These shadows and bright spots have played games with our eyes for thousands of years. Primitive people thought the shadows might be a person. That's why we've all heard about "the man in the Moon," even though we know there's no such thing. Until not so long ago, some people thought the Moon was a world like Earth. Of course, we now know that this is not true, either. Even in ancient times, there were tales about trips to the Moon. Thanks to our modern science and our old-fashioned curiosity, these tales have come true.

Over the years, people have seen many faces in the Moon's surface. Upper: This is how an artist imagines the Moon when the Sun's light reveals only one-quarter of the surface. Lower: Can you imagine shadows and light forming this jolly face when the Sun's light falls full on the Moon?

Discovering the Moon

In ancient times, people had to look at the Moon with their eyes only. Then, in 1609, an Italian scientist named Galileo built a telescope to make things look larger and nearer. The first thing he did was use it to look at the Moon. At once he saw it was a world. There were mountain ranges, and there were craters. A few craters had bright streaks coming out all around. The shadows on the Moon turned out to be flat, dark areas, and Galileo thought they might be seas of water. It turned out that they weren't, however. There is no water on the Moon — no air, either.

© National Geographic, Jean-Leon Huens

Galileo Galilei (1564-1642) argues with Church officials over his ideas about the skies. What Galileo saw through his telescope was so different from what people believed that they must have thought there was a problem with his scope!

© Dennis Milon

This is a photograph of the Moon taken in 1960 from Houston, Texas. Although it would not have been quite so clear to Galileo, imagine his pleasure at seeing this surface through his telescope!

The crater Langrenus. To get from one side to the other, you would have to walk about 85 miles (137 km). This is how Langrenus looked on December 24, 1968, from the Apollo 8 spacecraft orbiting the Moon.

NASA

The moon's craters — look out above!

The craters and "seas" on the Moon were caused by meteorites bombarding the Moon's surface. Most of these strikes occurred in the early days of the Moon. But meteor strikes may even have happened in recent times. On June 25, 1178, five monks in Canterbury, England, recorded that "a flaming torch sprang up, spewing out fire, hot coals, and sparks" from the edge of the Moon. We think that a meteorite must have struck the Moon just at the edge of the far side. There's simply no way of predicting when a large object might strike the Moon — or the Earth.

The Moon's Changing Face

Is moonlight really light that comes from the Moon? We know that it is not. The truth is, the light we see when we look at the Moon is sunlight that shines on the Moon's surface. The Moon moves around the Earth, and as it does, different parts of it are lit by the Sun. When it is on the opposite side of the Earth from the Sun, the side we see is all lit. We call this view the "full Moon." When it is on the side of the Earth that is near the Sun, the lit side is away from us and we don't see the Moon. In between, the Moon is partly lit. These are the Moon's phases. It goes from full Moon back to full Moon in about a month. In ancient times, people used the Moon as a calendar to tell time.

New (or Crescent) Moon   First Quarter

The phases of the Moon as photographed from Earth.

© Tom Miller 1988

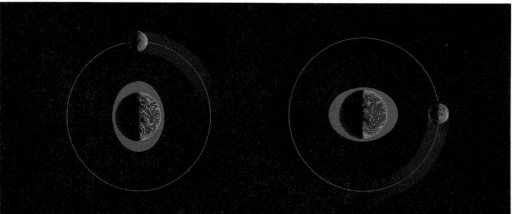

Tides are caused by the pull of the Moon's gravity on the surface of the Earth. Land is too firm to respond noticeably to the pull, but water stretches toward and away from the Moon because of gravity. In this diagram, the light blue egg-shaped area shows how the tides rise and fall around the world as the Moon orbits Earth.

ibbous Moon   Full Moon   Gibbous Moon   Last Quarter   Old Moon
Vaxing   Waning

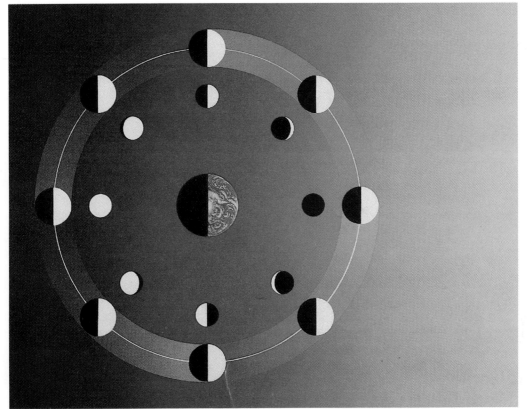

Images of the Moon reflecting sunlight as it circles Earth. The inner circle
shows us what the phases of the Moon look like from Earth as sunlight is
reflected on the Moon's surface. The outer circle shows us what the
Moon might look like from a point in space high above our North Pole.
From there, the Moon doesn't seem to go through phases at all.

Hide-'n'-Go-Seek

Usually, when the Moon travels about the sky and approaches the Sun's position, it goes a little bit above or below the Sun. Sometimes, though, it cuts right across the Sun and hides it for a while. This is called an eclipse. This was very frightening to ancient people who didn't know what was going on. They thought the Sun was dying! However, an eclipse of the Sun only lasts for a few minutes. On the other hand, sometimes, when the Moon is full and on the <u>opposite</u> side of the Earth from the Sun, it passes through the Earth's shadow. When the Earth's shadow falls on the bright side of the Moon, it makes the inner part of the Moon's surface dark. Then there is an eclipse of the <u>Moon</u>. That can last a couple of hours.

It is okay to watch an eclipse of the Moon, but staring into the Sun can hurt your eyes very badly. So you must never directly watch an eclipse of the Sun. And that means no telescopes or binoculars, either!

© Sally Bensusen 1988

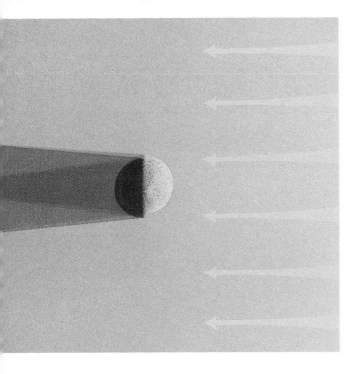

During a total eclipse of the Sun, the Moon blocks out the Sun's light from part of Earth. Within the smallest circle in this diagram, the sky would be quite dark and a person's view of the Sun would be that of the total eclipse. People within the outer circle would find daylight to be a strange kind of shadow and the Sun only partly eclipsed by the Moon. The photo above was taken from Earth during an actual eclipse. It gives a spectacular view of the Sun's corona.

During a Lunar eclipse, Earth comes between the Moon and the Sun and casts its shadow on the bright side of the Moon. The diagram illustrates this, and the photo shows it as it is happening. If you happen to be on the night side of Earth during a Lunar eclipse, you will be able to see the effects on the Moon as we slowly slide between Moon and Sun.

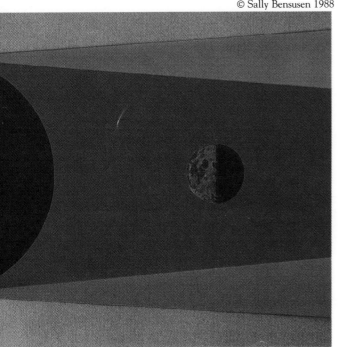

The Earth-Moon system. Compared to other natural satellites throughout the Solar system, our Moon is so big that we might ask whether Earth is more the Moon's <u>partner</u> than its <u>parent</u>. This photo was taken from a craft in Lunar orbit. It dramatically shows the blue Earth on the Moon's horizon.

The Double Planet Earth-Moon?

The Moon is quite large. It is 2,160 miles (3,456 km) across, a little over a quarter as wide as the Earth. The Moon's surface is as large as North and South America put together. The Moon isn't the only large satellite in our Solar system. Jupiter has four large satellites, two of them larger than the Moon. Saturn and Neptune each have a satellite larger than our Moon. However, Jupiter, Saturn, and Neptune are giant planets. It is amazing that a planet as small as Earth should have so large a satellite. Considering how small Earth is and how large the Moon is, Earth and Moon together are almost a double planet.

Will the real double planet please stand up?

The Moon has only 1/80 the mass of the Earth. Still, other planets have satellites with only 1/1,000 their own mass, or less. That is why Earth-Moon is considered to be a kind of double planet. But in 1978, it was discovered that the distant planet Pluto had a satellite. Pluto is a small world, even smaller than the Moon. Its satellite, Charon, is smaller still, but it is one-tenth the size of Pluto. Now it's Pluto-Charon that is the nearest thing to a double planet, especially now that astronomers believe Pluto and Charon are so close that they even share the same atmosphere! Earth-Moon is only in second place.

NASA

A composite photo of Earth-Moon as a double planet. Notice how close the Moon and Earth are in size. The view of the Moon is from Apollo 11 as it returned to Earth. The photo of Earth was taken from Apollo 17. You can see the weather systems in the Southern Hemisphere. Can you also make out Africa?

Exploring the Moon

We Earthlings have never been happy just to sit and stare at the Moon. Almost as soon as we began sending rockets into outer space in the 1950s, we aimed them in the direction of the Moon. In 1959, the Soviet Union sent a rocket past the Moon. It took pictures of the far side of the Moon, which we never see from Earth. Later in that year a Soviet rocket (with no people on board) landed on the Moon. US rockets were soon doing the same thing and were being put in orbit about the Moon. These Lunar Orbiters photographed all parts of the Moon close up. Scientists got to see all the details of the Moon's surface. Soon we would see even more.

NASA

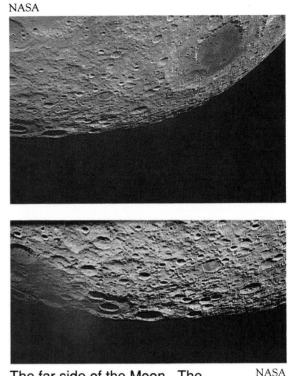

NASA

The far side of the Moon. The crew of Apollo 13 took these photos as they passed around the Moon. Upper: The large Lunar "sea" is called Mare Moscoviense after Moscow in the USSR. Lower: a view of the same area. The large crater on the horizon has the hefty name of International Astronomical Union Crater No. 221.

NASA

Models show an orbiting spacecraft circling the Moon and taking photographs less than 30 miles (48 km) from the Moon's surface.

Zond-3 probe. This Soviet spacecraft flew around the Moon and back to Earth.

Luna 3 probe. This Soviet research probe skimmed the Moon's surface to take photographs. In 1959, it produced the first photos we've seen of the dark side of the Moon.

Surveyor 6, a US soft-landing spacecraft. It allowed scientists to analyze the Moon by gathering material from the Lunar surface.

NASA

Stepping onto the Moon

Eventually, the Soviet Union and the United States began to put people on rockets. These people were called astronauts in the United States and cosmonauts in the Soviet Union. The US, in particular, decided to place astronauts on the Moon. During the 1960s, rockets flew closer and closer to the Moon. Finally, on July 20, 1969, the big moment arrived. Neil Armstrong set foot on the Moon and became the first human being to walk on another world. After that, five more rocket ships landed on the Moon, ran experiments there, and brought back Moon rocks for scientists to study. These rocks would give us a chance to look at the Moon in a whole new way. For starters, we found out for sure what scientists had suspected — the Moon was a completely dead world.

NASA

Top left: That's Astronaut David R. Scott of the Apollo 9 crew working outside his Earth-orbiting space-craft in 1969. The beautiful blue behind him is Earth. His partners are Russell L. Schweickart and James A. McDivitt.

Top right: Apollo 11 crew, left to right: Neil Armstrong, Michael Collins, Edwin ("Buzz") Aldrin. Collins orbited the Moon in Columbia, the command module, while Armstrong and Aldrin settled into the Moon dust in the Lunar module Eagle on July 20, 1969.

NASA

With no air on the Moon to blow it away, Buzz Aldrin's footprint could remain as it is shown here for billions of years. Of course, we Earthlings may have other plans for this stretch of Lunar real estate!

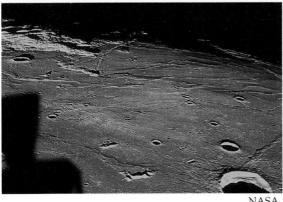

The Apollo 11 crew took this photo (left) of the Moon's surface in 1969. This is actually the approach to Apollo Landing Site 2 in the Sea of Tranquility.

Alan L. Bean, pilot for Apollo 12, gathers Lunar soil for research in 1969. Also featured in this photo is Charles Conrad, Jr., reflected in Bean's helmet.

Below: Apollo 15 Astronaut Dave Scott showed a worldwide TV audience that Galileo was right: "Gravity pulls all bodies equally, regardless of their weight." To do this, Scott dropped a hammer and a feather and watched as, with no atmosphere to alter their fall, they hit the surface of the Moon at the same time. Astronaut/artist Alan Bean watched on TV, too, and he painted this picture.

© Alan Bean 1986

Right: the US flag, held in a permanent "wave" by its wire frame, adds a dash of color to the Moonscape. Bundled up in his spacesuit, Apollo 15 Astronaut Jim Irwin also poses somewhat stiffly.

Where Did the Moon Come from?

So now we knew more about the Moon than ever before. But scientists still couldn't say for sure why Earth had such a large Moon. One theory went like this: When the Earth was formed, it spun so fast that a large piece of it split off. The trouble is, Earth never spun fast enough for this to happen. Or perhaps the Moon was an independent planet, and it was trapped by Earth's gravitation when it passed too close and was captured. That didn't seem likely, either. Or perhaps when Earth was formed, <u>two</u> worlds were formed. In that case, Earth and Moon should be made up of the same materials. The Moon rocks showed this was not so. The whole thing was a puzzle.

© William K. Hartmann

In this artist's conception, the Moon is shown having formed when it was much closer to the Earth than it is today. Here we also see a ring of leftover debris accompanying the Moon in its orbit around Earth.

NASA

Some unusual Lunar close-ups.

Upper right: an electron microscope's view of Lunar dust (Apollo 16).

Lower right: Moon rocks (Apollo 11).

Upper left: a farming experiment — soybean sample exposed to Lunar soil (Apollo 15).

NASA

NASA

The mystery of our two-faced Moon?

One side of the Moon always faces us. The other side always faces away from us. Once the Soviets and Americans had photographed the other side, scientists discovered that the two sides were quite different. The side that faces us has the large, flat dark areas we call "seas" (even though there is no water in them). The far side has only a few small seas and far more small craters. That would make it appear that meteor strikes and volcanic eruptions — the two main causes of craters — occurred at different rates on each side of the Moon. Why? We're not sure.

A New Theory

Then, just a few years ago, scientists thought about something that may have happened when Earth was first created and other worlds were coming into being. What if one, about a tenth as large as Earth, passed close to our world? It wasn't captured. Instead, it <u>hit</u> the Earth a glancing blow, knocked a piece off, and went on its way. Scientists set up a computer program that showed what would happen if such a world hit the Earth. The computer showed that something like the Moon would be formed out of Earth's outer layers but without the inner ones. That would explain why the Moon doesn't have the same makeup as Earth.

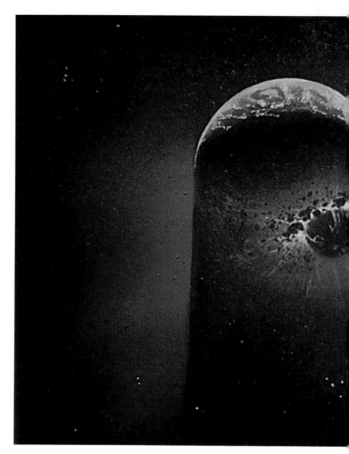

Lunar time vs. Solar time

Ancient peoples who used the Moon for a calendar measured their years in "lunar years." There would be 12 new Moons from one spring to the next. But that wasn't quite enough to fill a whole year. So every couple of years they would add a month and count 13 new Moons to the year. Later, people decided it was easier to make the months a bit longer so that there were always 12 months to a year. The date of Easter is still based on the old lunar calendar. That's why it keeps changing dates from year to year. Muslims also use the lunar months, but with only 12 months a year. That makes their year only 354 days long.

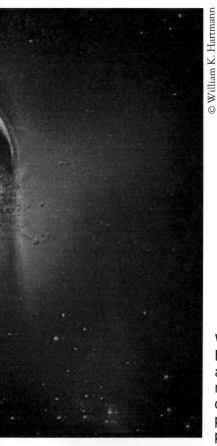

Was it a collision in Earth's formative years that blasted our Moon into orbit? Here are two views. Left: A giant asteroid about one-tenth Earth's size slams into our planet, blasting material out of Earth's outer layer. Below: Another planet collides with ours. After the collision, debris from the other planet spreads out in space and, thanks to its gravitational pull, eventually clumps together.

Our Next Frontier?

Is there any chance that people might one day work — or even live — on the Moon? It wouldn't be easy to try to live on the Moon. It's nothing at all like Earth. For one thing, the surface gravity is only one-sixth that of Earth. Also, there is no air or water on the Moon. And the Moon turns so slowly that the day and night are each two weeks long. During the day, the temperature rises to higher than the boiling point of water. During the night, the temperature gets colder than Antarctica. And without an atmosphere, there is nothing to filter out the radiation in sunlight, or to burn up the little meteorites that are always striking. There is also no magnetic field to turn away cosmic rays.

© LPI 1985, Pat Rawlings

A child and an adult survey the scene of this Lunar base. Here is where the mining of our Moon's natural resources takes place. The six-mile-long (10-km-long) mass driver shown here would provide the boost needed to power payloads off the Moon.

Another artist's conception of a Lunar base — a colony where people live, work, and play as Lunar residents. In both these views of life on the Moon, people must live within the totally artificial environments of their buildings, vehicles, and spacesuits. Such a setting might help prepare future "space people" for their lives as permanent settlers of the cosmos.

A Lunar magnetic field — yes or no?

The Earth has a magnetic field, but the Moon has not. Earth has a large, hot core of liquid iron that swirls as the Earth rotates. That produces the magnetic field. The Moon is less dense than Earth, so it must have only a small core of heavy iron, perhaps none at all. Even if it had a metal core, the Moon isn't large enough to keep it hot and liquid. Still, the Moon's rocks show signs that they were affected by magnetism. Could the Moon in its early days have had a hotter center than now? Could it have had a magnetic field that would have affected its early history? We aren't sure.

Living on the Moon

Does living on the Moon interest you? It sounds as if it could be a tough life. But it might still be possible to live on the Moon, if people stayed a few yards <u>under</u> the surface. There, the temperature is always mild, and people would be protected from the Sun's radiation, from meteorites, and even from cosmic rays. People on the Moon could do valuable work by setting up mining stations. The Moon's surface could yield all the construction metals, as well as oxygen, glass, and concrete. People could make building parts that would be fired into space easily because of the Moon's low gravity. These parts would be used to build places where people could live and work out in space.

The tides — are they wearing Earth down?

Because the tides rise and fall, there is friction of water against the shallow sea-bottoms. The friction consumes some of the energy of the Earth's rotation. As a result, our days are slowly growing longer and the Moon is slowly moving farther away. These changes are so slow that in all history they haven't been noticeable. In very old times, however, the Moon was closer to Earth, the day was shorter, and the tides were higher. How did this affect the development of life? Did the higher tides make it easier for sea life to crawl out on land? We just don't know.

© Doug McLeod 1988

A robot craft operated by a cosmic construction worker puts layers of insulation made from Lunar soil on an immense colony between Earth and the Moon. Other craft approach the docking area, which is the light spot on the colony's "roof," and tube-like structures on the top are what dozens of human workers call home. These details give you an idea of the size of this human habitat in space.

Science from the Moon

Someday, we may mine the Moon for building materials and energy resources. But there are other uses for the Moon, and we must be careful not to disturb the Moon too much. The Moon is smaller than Earth, and it has changed less since the early days of the Solar system. This means that we can study the first billion years of the Solar system easier on the Moon than on Earth. Then, too, on the far side of the Moon we can set up large light-telescopes and radio-telescopes. There would be no atmosphere to interfere, no Earthly lights or radio signals. We could see farther and more clearly into deep space and learn about the very early days of the Universe.

Who knows what mysteries we may uncover about our Earth — and our Universe — now that we have walked on the Moon?

Imagine what it would be like to look at space from a site on the Moon. In this artist's conception, Soviet and US workers break ground for a huge multi-mirror telescope on the far side of the Moon. In the background is a radio/optical observatory.

Imagine seas on a terraformed Moon! By creating an atmosphere on the Moon, we could capture sunlight and turn the Moon into a celestial tourist trap. This would be fun, but many scientists feel it is more important to keep the Moon pretty much as it is. Then we could use it to help us better understand Earth and the cosmos.

Upper: An artificial satellite hovers above the Lunar seas.

Right: Moon tourists have discovered the pleasures of this Moon beach.

Fact File: The Moon's Craters

Today, thanks to Lunar probes and piloted missions, we have seen the Moon's craters close up, as well as something never before seen by humans — the far side, which always faces <u>away</u> from Earth. On these two pages, you can examine two interesting questions about the Moon's craters: 1) How were they formed? 2) Why are the craters on the far side so different from the ones on the near side?

How Were the Moon's Craters Formed?

By the Impact of Meteorites	By Volcanic Action

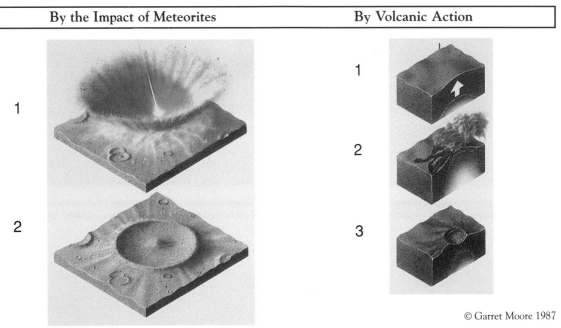

© Garret Moore 1987

By the Impact of Meteorites

1. Meteorite strikes the Moon's surface, sending out a shock wave that gouges out deep hole, and throwing up a cone-shaped curtain of boulders and other debris that fall back to surface.

2. The boulders create several smaller craters around the first one, and the finer debris settles into a blanket.

Comments:
• Upon impact meteorite is consumed, or absorbed, into its crater.
• Matter at center of impact "rebounds," just as a drop in a pool of water would, and freezes.
• Thin lines, or filaments, emerge as blanket of dust settles. Pattern of lines called <u>rays</u> extends outward from crater.
• Most of Moon's craters have been formed by impact of meteorites.

By Volcanic Action

1. Portion of surface forced upward by melted rock and gases from within Moon's interior.

2. Eruptions of gas and lava through Lunar surface and into sky above. Pressure from below now eased.

3. Collapse of surface into a crater.

Comments:
• Signs of volcanic crater differ from those of meteorite crater.
• No rays, smaller craters nearby, or "peak" at center of volcanic crater.
• Volcanic craters are a sign that Moon once had a very active, hot inner region.
• Unlikely there would be any current volcanic activity on Moon — just some possible shifting or adjusting of Moon's surface. These shifts might give rise to an occasional volcanic "burp" of trapped gas.

Comparing Craters — The Near Side vs. the Far Side

Near Side

As these photos illustrate, the Moon's near side has fewer craters of the type found on the far side. But it has more of the maria, or "seas," that show up as large dark areas. The maria are actually the result of volcanic activity that covered ancient meteorite-impact craters with flowing lava. Why did so much more volcanic activity occur on the near side, and why did so many more meteorites seem to have struck the far side? Scientists aren't sure.

Perhaps more meteor strikes occurred on the far side because Earth partly "blocked" the near side from meteoroids. And, perhaps, there were more volcanic eruptions on the near side because of the pull of Earth's gravity on the gases and melted rock below the Moon's surface. No one knows for sure.

Far Side

More Books About Our Moon

Here are more books about the Moon. If you are interested in them, check your library or bookstore.

A Close Look at the Moon. Taylor (Dodd, Mead)
All About the Moon. Adler (Troll)
The First Travel Guide to the Moon. Blumberg (Scholastic)
The Moon. Barrett (Franklin Watts)
Moon Flights. Fradin (Childrens Press)

Places to Visit

You can explore our Moon and other places in the Universe without leaving Earth. Here are some museums and centers where you can find a variety of space exhibits.

Hayden Planetarium
Museum of Science
Boston, Massachusetts

Henry Crown Science Center
Museum of Science and Industry
Chicago, Illinois

Seneca College Planetarium
North York, Ontario

NASA Lewis Research Center
Cleveland, Ohio

Dryden Flight Research Center
Edwards, California

Edmonton Space Sciences Centre
Edmonton, Alberta

For More Information About Our Moon

Here are some people you can write away to for more information about the Moon. Be sure to tell them exactly what you want to know about or see. And include your full name and address so they can write back to you.

For information about the Moon:
STARDATE
MacDonald Observatory
Austin, Texas 78712

Space Communications Branch
Ministry of State for Science and Technology
240 Sparks St., C. D. Howe Bldg.
Ottawa, Ontario K1A 1A1
Canada

For catalogs of posters, slides, and other astronomy materials:
Selectory Sales
Astronomical Society of the Pacific
1290 24th Avenue
San Francisco, California 94122

About missions to the Moon:
NASA Kennedy Space Center
Educational Services Office
Kennedy Space Center, Florida 32899

Glossary

Armstrong, Neil: the first person to touch the Moon's surface (1969).

astronauts: men and women who travel in space.

atmosphere: the gases that surround some planets; our atmosphere consists of oxygen and other gases.

billion: the number represented by 1 followed by nine zeroes — 1,000,000,000. In some countries, such as the United Kingdom (Great Britain), this number is called "a thousand million." In these countries, one billion would then be represented by 1 followed by 12 zeroes — 1,000,000,000,000: a million million.

craters: holes caused by meteor strikes or volcanic explosions.

eclipse: when one body crosses through the shadow of another. During a Solar eclipse, parts of the Earth are in the shadow of the Moon as the Moon cuts right across the Sun and hides it for a period of time.

eclipse of the Moon: what occurs when the Moon is full and on the opposite side of Earth from the Sun, and then passes through Earth's shadow.

full Moon: what we call the Moon when it is on the opposite side of Earth from the Sun so that it is lit in its entirety.

Galileo: an Italian scientist who made a telescope and got the first clear view of the Moon's surface.

Lunar Orbiters: vehicles that flew to the Moon and photographed all parts of it, including the previously unseen far side.

lunar years: the basis for ancient calendars. There would be 12 new Moons from one spring to the next.

mare: (pronounced "MAH-ray") the Latin word for "sea." The plural of "mare" is "maria" (pronounced "mah-REE-ah"). People once thought the Moon's flat dark areas contained water, and so they called these areas maria. Today we know they were caused by volcanic eruptions producing lava flows.

Moon: Earth's only satellite. It is about 250,000 miles (400,000 km) from us.

phases: the periods when the Moon is partly lit by the Sun. It takes about one month to progress from full Moon to full Moon.

Pluto-Charon: the combination of planet and moon that is the nearest thing to a double planet. Astronomers believe that Pluto and Charon may even share the same atmosphere.

radio telescope: an instrument that uses a radio receiver and antenna to both see into space and listen for messages from space.

"seas": the name for the flat dark areas on the Moon or Mars, even though they are completely waterless. Any one of these "seas" is actually called a "mare" (pronounced "MAH-ray").

terraform: to make another world Earth-like by giving it qualities that are, as far as we know, special to Earth, such as an atmosphere and water. "Terra" is Latin for "earth."

Index

The publishers wish to thank the following for permission to reproduce copyright material: front cover, © Frank Zullo 1985; p. 4, Harvard College Observatory; pp. 5 (both), 10-11 (upper and lower), © Sally Bensusen 1988; p. 6, © National Geographic, Jean-Leon Huens; p. 7 (upper), © Dennis Milon; pp. 7 (lower), 12, 13 (both), 14 (all), 15 (lower), 16 (all), 17 (upper left, upper right, and lower right), 19 (all), 29 (lower), courtesy of NASA; pp. 8-9 (all upper), 29 (upper), Lick Observatory; pp. 8 (lower), 9 (lower), © Tom Miller 1988; p. 11 (upper and lower right), © George East; p. 15 (upper left and upper right), Oberg Archives; p. 17 (lower left), © Alan Bean 1986; pp. 18, 20-21 (upper), © William K. Hartmann; p. 21 (lower), © Ron Miller; p. 22, courtesy Lunar & Planetary Institute-© 1985, Pat Rawlings; p. 23, © Mark Paternostro 1978; p. 24, © Doug McLeod 1988; p. 26, © Paul DiMare 1986; p. 27 (both), © David Hardy; p. 28 (both), © Garret Moore 1987.

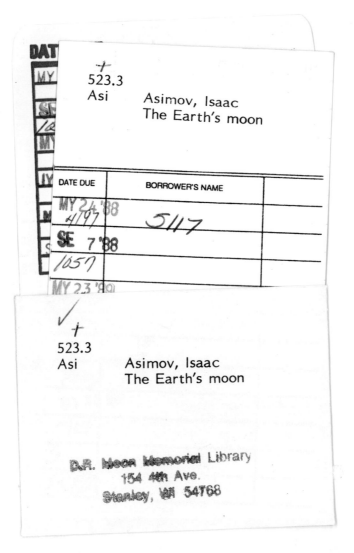

+
523.3
Asi   Asimov, Isaac
           The Earth's moon

DATE DUE	BORROWER'S NAME	
MY 24 '88 4/97	5117	
SE 7 '88		
1057		
MY 23 '9?		

✓ +
523.3
Asi   Asimov, Isaac
           The Earth's moon